COLLECTION OF THE 19TH ASIA-PACIFIC INTERIOR DESIGN AWARDS

第十九届亚太区室内设计大奖参赛作品选

FOOD SPACE
用餐空间

翟东晓 / 深圳市创福美图文化发展有限公司　编著

大连理工大学出版社

图书在版编目(CIP)数据

　　第十九届亚太区室内设计大奖参赛作品选.用餐空间
:英汉对照/翟东晓,深圳市创福美图文化发展有限公
司编著.—大连:大连理工大学出版社,2012.12
　　ISBN 978-7-5611-7355-8

　　Ⅰ.①第… Ⅱ.①翟…②深… Ⅲ.①室内装饰设计
—亚太地区—图集②餐馆—室内装饰设计—亚太地区—图
集 Ⅳ.① TU238-64 ② TU247.3-64

　　中国版本图书馆CIP数据核字(2012)第233872号

出版发行：大连理工大学出版社
　　　　　（地址：大连市软件园路80号 邮编：116023）
印　　刷：利丰雅高印刷（深圳）有限公司
幅面尺寸：242mm×263mm
印　　张：17
插　　页：4
出版时间：2012年12月第1版
印刷时间：2012年12月第1次印刷
责任编辑：裘美倩
责任校对：王丹丹
文字翻译：梁先桃
装帧设计：刘竞华

ISBN 978-7-5611-7355-8
定　　价：240.00元

电话：0411-84708842
传真：0411-84701466
邮购：0411-84703636
E-mail:designbooks_dutp@yahoo.cn
URL:http://www.dutp.cn
设计书店全国联销：www.designbook.cn

如有质量问题请联系出版中心：（0411）84709246 84709043

The exclusive distributorship in Taiwan China is offered to ArchiHeart Corporation. Any infringement shall be subject to penalties.
中国台湾地区独家经销权委托给ArchiHeart Corporation(心空间文化事业有限公司)，侵权必究。

PREFACE
前 言

Kinney Chan
Chairman of
Hong Kong Interior
Design Association

The year of 2011 was the 20th anniversary of the HKIDA, and 2012 is the 20th edition of APIDA. I am happy to report that in recent years, APIDA become more and more international, and now includes entries from Asia and all across the world. It is now one of the most recognized international interior design awards, and one of the most reputable awards of its kind in the world.

On behalf of the HKIDA, I would like to extend my deepest thanks to all the jury members for all their dedication and efforts, and to all the sponsors for making it possible for us to realize the aims and mission of APIDA. Also I want to thank all the designers who entered their works to this competition and made it so fantastic.

2012 is the Hong Kong Design Year, and the Hong Kong government will support a series of mega events to showcase Hong Kong's strength as a regional design hub, foster the community's interest in design and celebrate design excellence. These events include international design forums, regional exchange symposia and exhibitions. The HKIDA will contribute to the Hong Kong Design Year by using APIDA as a platform to foster interior design excellence, promote international exchange and showcase the most outstanding interior designs in the Asia-Pacific region. I hope interior designers from Hong Kong China and the surrounding regions will continue to express their originality, creativity and passion for design through their works, and contribute to making Hong Kong China the design capital of Asia.

Po Po Leung
Chairlady of APIDA 2011

Last year HKIDA was celebrating its 20th anniversary, and this also happened to be APIDA's 19th edition. From 2010, APIDA underwent a re-branding and re-designing of its image, and you would have noticed changes in our promotion materials and award statues. We also made our promotion strategies more international, and have invited official media partners in different regions such as Mainland China, Taiwan China, Hong Kong China, Japan, and Malaysia to join us in publicizing the event and increase APIDA's presence in these places.

However, the one thing that remains unchanged is APIDA's mission of celebrating excellence in interior design and raising the professional standards and conduct of the industry in the Asia-Pacific region.

For the 19th APIDA, we are happy to receive nearly 600 entries from Hong Kong China, Mainland China, Macau China, Taiwan China, Singapore, Malaysia, Thailand, the Philippines, Indonesia, Korea, Japan, Australia and New Zealand. Overall the standard of the entries is spectacular, which made the job of our jury members very difficult indeed. On behalf of the HKIDA, I offer my deepest gratitude to our jury members for their time, hard work, dedication and support. Without you this competition would not have been possible.

With its 19-year history, APIDA has become not only the most well-known interior design award in the region, but also the most widely respected and professional event of its kind. The award itself is a symbol of excellence coveted by interior designers from across Asia and beyond. I congratulate the design teams behind all of this year's entries, and thank you for using APIDA as a platform for sharing and exchanging ideas. I myself have learnt a great deal, from your works, and I hope in the future you will, continue to strive for excellence and keep on creating interior environments that benefit your clients, users and society as a whole. Thank you very much.

目录 Contents

6	眉州小吃北苑店	108	DIM SUM BAR
10	Green@Hotel ICON	112	Envie
16	MIST	116	Orange Terrace
20	The Loft	120	Gion NITI
26	上堡藏馆	124	亚洲土地
32	Japanese Restaurant Kouraibashi KITCHO	128	叙品咖啡厅
36	八沪曲西餐厅	132	北京辉哥火锅
40	厦门宴遇时尚餐厅	136	和宴
44	伴山惠馆	140	南京紫轩餐饮会所
48	长沙饭怕鱼连锁餐饮湖大店	144	Spice Garden
52	食品厂	148	Derby Restaurant & Bar
56	八方馔养生餐厅	152	苏州酒田日本料理
60	阿度餐厅	156	OHANA
64	隐庐私厨	160	阿林鲍鱼延安店
68	寻常故事	164	御膳皇庭中餐厅
72	香舍会	168	Dolar Shop
76	晶宴中和会馆	172	Hannanshan Korean BBQ
80	Yakoya Kelapa Gading	176	肴肴领先饮食会所
84	宴遇·乡水谣	180	上海桃花源餐厅
88	House of Wine	184	港盈轩餐厅
92	唐华乐府	188	Café Regal
96	Pollen Street Social	192	爱尚渔香主题餐厅
100	Yarra Lane Precinct	196	Fish & Co
104	Mercedes-Benz Connection		

经典国际设计机构（亚洲）有限公司　/　王砚晨、李向宁　/　马艳

中国北京市朝阳区北苑家园易仕达广场世代百货 5 楼 / 360m²

眉州小吃北苑店

The designers hope to use the most basic materials and the simplest techniques to solve the conflict of various elements, making the Meizhou natural spirit co-exist with the modern urban space. The whole restaurant adopts simple materials and smooth forms to create a comfortable atmosphere. The fluidity and dynamic of the space is running through the restaurant with people stream, forming a harmonious space image.

设计师期望运用最本质的材料和最洗练的手法来解决各种元素的碰撞，让眉州的自然气息与时尚都市空间交融共生。整个餐厅采用单纯的材质和流畅的形体来营造一种轻松的氛围。空间的流动性和动感随同人流在餐厅中穿行，形成和谐的空间意象。

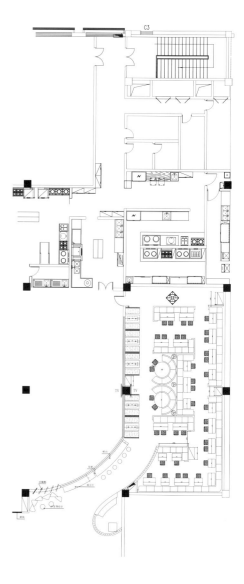

CL3 Architects Limited / William Lim / Joey Wan, Jane Arnett, Carmen Tang, Kinson Tam, Patrick Blanc

Hong Kong, China / 260m²

Green@Hotel ICON

Located at the bottom of a glass atrium, its name reflects a large vertical green wall. Curved high back seating creates intimate spaces in this large volume. A service bar with a tall wine storage unit brings in a vintage Hong Kong sliding gates motif, which also repeats at the main entrance portal. A high table incorporating the word "ICON" becomes the focus of this popular nighttime hangout.

这间酒吧位于一个玻璃房的底部，它的名字来自一面巨大的绿色垂直墙壁。弧形高背椅为这个大空间营造了一种亲密性。带有高大的葡萄酒储柜的服务吧台采用香港古老的推拉门模式，这一模式也应用到主入口处。绘有"ICON"标识的高台成为这间大众化夜场的焦点。

Green@Hotel ICON

Green@Hotel ICON

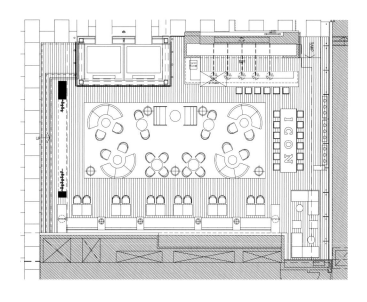

Food Space 15

福州林开新室内设计有限公司 / 林开新 / 陈青青

中国福建省福州市 / 100m²

MIST

The project is a fashionable commercial space which mainly sells desserts and coffee. The owners of this place are two young beautiful women, whose fashion dress and elegant class is just the inspiration of the design thought. The design uses arc line structure to activate the spatial ambience, and uses the white and black color to define the fashion theme, whilst the use of LED lighting makes the ceiling which is covered with PVC hose vivid and interesting.

本案是主营甜品、咖啡的时尚商业空间。业主是两位年轻的美貌女性，其时尚的着装、优雅的气质正是本案的设计思路来源。设计用弧线结构来活跃空间气氛，用白与黑的色彩组合来诠释时尚的主题，而LED照明使满铺PVC软管的吊顶生动有趣。

MIST

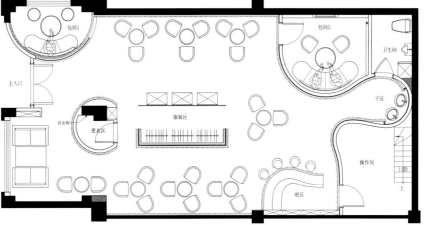

Joey Ho Design Limited / Joey Ho / Althea Lee, Ray Lau

Hong Kong, China / 365m²

The Loft

Loft-style is applied in this restaurant for bringing diners an idyllically homely experience. Raw materials and the simple contrast of the interior red and white frame system visually generate fusion of old and new elements. This minimalist layout evokes an unfinished yet chic and airy ambience. The semi-framed dining area provides privacy in this open-concept space. The designer has transformed the existing niche space of the restaurant into a loft with light-filled and domestic scene.

餐厅采用阁楼式风格，带给用餐者一种温馨的居家体验。纯天然的材料以及室内的红色和架构的白色的简单对比，制造出视觉上新与旧的融合。这种简单抽象的布局营造出一种未经雕饰却雅致大气的氛围。在这间开放式理念的空间中，半围式的用餐区提供了一种私隐性。设计师将餐厅现有的壁龛空间转变成为一间光线充足且拥有内景的阁楼。

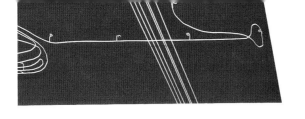

The Loft

The Loft

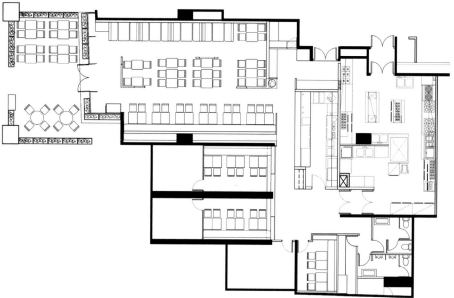

香港格瑞龙国际设计有限公司 / 曾建龙

中国浙江省温州市 / 94m²

上堡藏馆

The project mainly collects purple enameled pottery, which is with the cultural atmosphere of tea ceremony. The use of contemporary orient design language demonstrates the space, zoning two functional areas, public hall area and the two compartments. The design shapes the space structure by lines and surfaces, so as to show the art taste of the space. The space uses the black and white color as the main tone, better expressing the texture of the collections. The whole space is full of thick orient culture, stressing the character of meeting friends through tea.

本案以收藏紫砂壶为主，带有茶道文化的气氛。运用当代东方设计语言进行空间诠释，划分了两个功能区、公共大厅展示区以及两个包间。设计通过线、面的关系来进行空间结构塑造，从而传递了空间的艺术气息。空间以黑白色为主色调，更好地表现了收藏品的质感。整体空间东方文化浓重，突出以茶会友的特色。

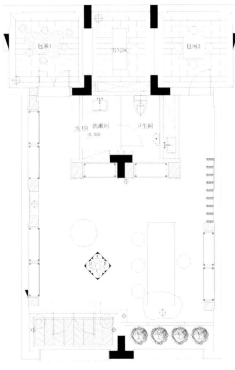

Osaka, Japan / 237.3m²

Japanese Restaurant Kouraibashi KITCHO

Established in Osaka eighty years ago, the restaurant uses traditional design and materials under the Japanese concept of harmony and Japanese Tea Ceremony dining. The walls, ceiling, and even table surface are covered with Japanese paper, and the use of hammered brass plate for doors, walls and space, all reflects the characteristic of Japanese tradition.

餐厅建于八十年前的大阪，受日本和谐理念及日本茶文化礼仪的影响，采用了传统设计和材料。墙面，天花，甚至桌面都铺上了日本图纸，门上、墙上及空间里黄铜镀板的应用，无不流露出传统的日本特色。

Japanese Restaurant Kouraibashi KITCHO

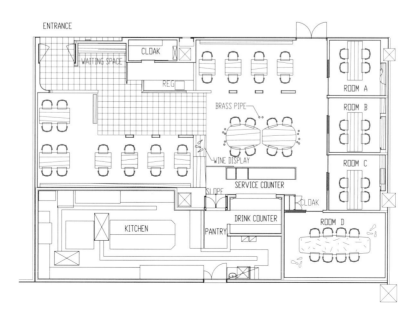

山隐建筑室内装修设计有限公司　/　何武贤　/　吕嫦谋

Taipei, Taiwan, China

八沪曲西餐厅

This is a western restaurant of a club. Except the existing Shanghai style of the club, some modern elements are specially fused in the design, such as an acrylic cage for birds, polished stainless steel images cut by laser, mirror mosaics, a shining wall made up of a series of parallel illuminated glass plates, and blue LED lights. The spirit of a modern lifestyle blends in the original mellow, elegant atmosphere of old times.

这是一个会馆中的西餐厅。设计除了沿袭会馆既有的上海风格外，还特别安排了一些现代元素融入其中，如树脂材质的鸟笼、镜面不锈钢的切割图像、镜面马赛克、玻璃光墙及蓝色LED灯光。在原有温润典雅的古老氛围中，注入了时尚生活的现代精神。

八沪曲西餐厅

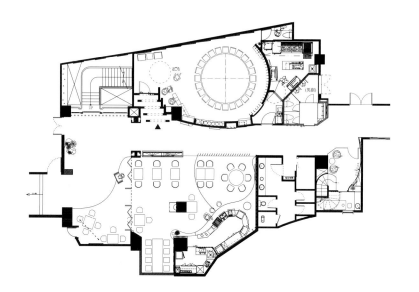

厦门东方设计装修工程有限公司 / 吴伟宏 / 宏盟东方室内设计机构

中国福建省厦门市莲坂新景广场 2 层 / 2500m²

厦门宴遇时尚餐厅

The project is positioned as trendy restaurants. The design, with the style of mix and match elements, is an innovation of classical and modern, Western and Eastern culture. In the aspects of selection of materials, the extensive use of environmentally friendly paint, partial stainless steel and copper and the audience with LED saving control system, creates a comfortable dining environment.

该案定位为时尚餐厅。设计采用混搭的风格元素，是一次古典和现代、西方与东方文化的创新。在选材上大量采用环保涂料，局部采用不锈钢及铜片，全场采用 LED 节能控制系统，营造一个舒适的用餐环境。

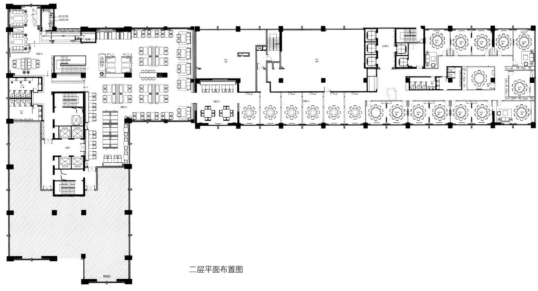

二层平面布置图

无锡市上瑞元筑设计制作有限公司 / 冯嘉云 / 徐小安

中国江苏省无锡市北塘区惠山直街 186 号惠山古镇内 / 1500m²

伴山惠馆

The interior design focuses on the spatial effect. First, it becomes harmonious with external environment after classical elements are employed on its door and signboards, which are in conformity with its image of "landscape restaurant". Second, combining typical Chinese elements and vogue international colors, the Hall shows good taste of upper class. Finally the deeply colored solid wood materials with veins, the potteries and the sculptures jointly create a significant space where dining and communication would become more enjoyable, which is quite fit for private family gathering or business banquet.

室内设计注重空间效果。首先是在门、店招等外部元素上渲染古典氛围，取得与周边环境的和谐呼应，建立"景观饭店"概念。其次是通过典型中式元素与国际化主流色的混搭塑造属于高端人群的品位感。最后通过深色纹理实木材料、浅釉陶瓷、雕刻等细节打造出隽永而意味深长的空间，增进饮食、交流，适合偏重私密性的家庭聚会与商务宴请。

伴山惠馆

长沙水木言室内设计事务所 / 梁宁健 / 陈勇苗、贺俊

中国湖南省长沙市河西大学城渔湾码头商业广场 / 3000m²

长沙饭怕鱼连锁餐饮湖大店

As a chain restaurant specializing in freshwater fish cooking, the Hunan University Branch of Changsha Rice Afraid Fish Restaurant is situated in Changsha Higher Education Mega Center, with millenary Yuelu Academy on the north and Xiang River on the east. Therefore, on the application of design elements, the designers give attention to the characteristics of being close to mountain, water and culture as well as combine the life elements related to fish, aiming to create a dining space with distinctive cultural individuality.

长沙饭怕鱼湖大店坐落于长沙大学城内，北靠千年岳麓书院，东临湘江，是以出品河鱼为主的连锁餐饮，因此在设计元素运用上以亲山亲水亲文化的特征，再结合与鱼相关的生活元素，旨在打造文化个性鲜明的餐饮空间。

长沙饭怕鱼连锁餐饮湖大店

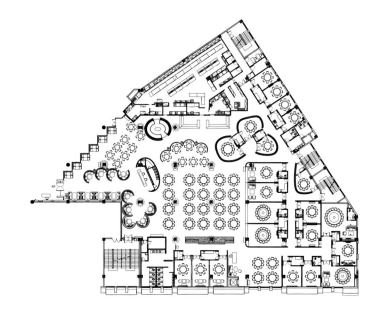

厦门三佰舍室内设计顾问有限公司　/　方令加　/　李少东

中国福建省厦门后江埭路 29 号 / 700m²

食品厂

The project is the first health restaurant dealing mainly the health food of "Sea Cucumber", following the order of nature of birth in the spring, growth in the summer, harvest in the autumn, collection in the winter, and customizing the unique health cuisines that fit for different people. The whole space design is full of original savour, which is no delicate painted walls but the original "rough blank" of red bricks, as well as is no magnificent ornament ceilings but the bamboo screen and the antique logs are partially hidden. The oil paintings on the wall are peculiar; the chairs and dining-tables are simple and elegant, which are just like nature itself without deliberate treatment. The tatami and tea tables in the boxes provide a tranquil ambience of Zen spirit for people to enjoy after dinner.

本案是厦门第一家主打"辽参"养生菜的养生餐厅，遵循天人合一春生夏长秋收冬藏的自然规律，定制出适合不同人群的独特养生菜系。整个空间设计充满原始味道，没有精致粉刷的墙壁，只有原始的红砖"粗坯"，没有华丽的修饰吊顶，只有若隐若现的竹屏和复古圆木。墙上的油画别具一格；座椅和餐桌素雅古朴，不需刻意雕琢，仿佛浑然天成。包厢内设有榻榻米和茶几，用餐之余更是一种享受，一种如禅宗文化般的静谧。

厦门宽品设计顾问有限公司 / 李泷 / 刘可华、张坚

中国福建省厦门市金桥路 / 900m²

八方馔养生餐厅

The project is a characteristic restaurant which makes regimen food and provides Eight Cuisines. Based on the simple Chinese style, the design skillful fuses the Chinese classical elements in the modern space. The black-gold color traditional Chinese realistic painting in the hall, the wall with laconic ample texture, the fabrics partition with the simple Zen soul and hemp paper droplight with freehand flower and bird, all are showing deep orient verve. The space uses the rice color as the theme color, together with the gray color, showing a steady, low-key and modern fashion feeling.

本案是一间以制作养生美食、拥有八大菜系为特色的餐厅。设计以简约中式为基调，将中式古典元素巧妙地融入现代空间中。大堂富有中国风的黑金色工笔漆画，简洁丰富的肌理墙面，素雅禅意的布艺隔断，写意花鸟的麻纸吊灯，无一不渗透着浓浓的东方意韵。空间以米色为主题色，搭配灰色作为过渡，显得沉稳、低调又不失现代时尚感。

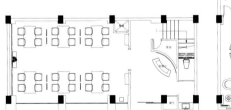
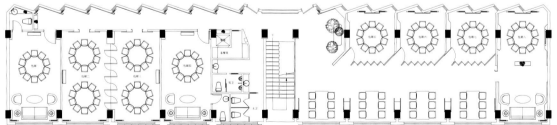

中国福建省厦门湖里区 SM 广场 2 期 / 346m²

阿度餐厅

The whole restaurant space layout is natural and free. The black and white color is the main color tone, together with the blue and green for the local embellishment. In order to have a pleasant dining environment, the designer adds a lot of interesting elements in the restaurant. The entrance uses different white gradient changes to increase the dimensions of the single tone design, combined with the irregular oval decoration, which boasts personality and strong visual impact. The large chandelier and birdcage wall lamp are exquisite for the space. The design utilizes the light belt and shine light to foil the warm dining atmosphere.

整个餐厅空间的布局自然顺畅。色调以黑色、白色为主,局部用蓝色与绿色作点缀。为了有个愉悦的就餐环境,设计师在餐厅设计中增加了许多趣味性。入口运用不同的白色梯度变化来增加单色调设计的维度,加入不规则椭圆形装饰,个性而又具有视觉冲击力。大型吊灯与墙壁鸟笼壁灯将空间装点得极为别致。设计利用灯带与射灯烘托餐厅温馨的就餐氛围。

中国湖北省武汉市武昌区东湖小路 / 800m²

隐庐私厨

The design focuses on making this restaurant a real space for acquiring quiet from the hustling city. The entry of the restaurant is reconstructed to lapse from the street. The interior selects fuscous terrazzo floor, mild wooden wall and cold-colored handmade tiles to add nostalgic touch. The restaurant is permeated with a kind of oriental introversive refinement after accomplishment. As a poem goes: The most capable people stay in worldly life while those seeking for peaceful mind stay in Hidden Cottage which is remote and quiet, the restaurant is therefore named "Hidden Cottage".

设计的重点是使餐厅能够"闹中取静"。重新规划餐厅入口使其背离街面。室内选用平静的水磨石地面、温和的实木墙板以及颓废的手工瓷砖来营造怀旧感。完成后的餐厅弥漫着一种东方内敛的精致。所谓"大隐隐于市",故餐厅取名"隐庐"。

隐庐私厨

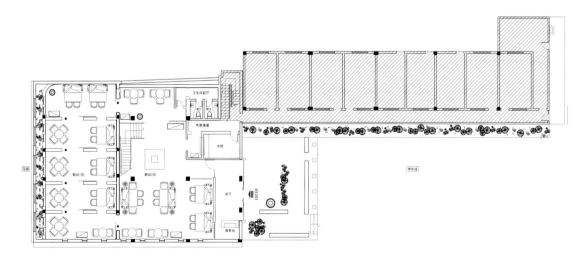

Food Space

中国福建省福州市 / 580m²

寻常故事

Aimed at creating a simple new oriental original space, the design adopts rattan products as the main material. The rattan chandelier of cloud decoration acts as the imaginary blue sky and white cloud. The wall of China red voile brings warmness to every visitor. The soft texture gives limitless kindness to people. The red above the partitions is the extension of the colors.

本案旨在打造质朴的新东方原生态空间，所以设计采用藤制品作为主要材料。藤制品云彩装置吊灯营造了意念中的蓝天白云。通道的中国红纱墙温暖着每个到访者。柔软的质感带给人无限的温情。隔断上方的红是色彩的延续。

Food Space 71

深圳市大象空间展览展示设计有限公司 / 张达利、易锦鸿、鲁晓勇

中国广东省深圳市侨城东路天鹅堡会所 / 1800m²

香舍会

Fronting toward the lake, the project is a detached two-story building with beautiful scene. The design creates a top fashion club in modern culture by the use of interior design style of decoration doctrine. At the corner, the well-known pictures, sculptures and art furnishing project a unique elevated taste. The Chinese and Western food served by the cooking expert, the private cigar room, the special red wine cellar, the store of works of art and high-end club of membership, all present fine operation philosophy and humanity care.

本案是一座独立的2层建筑，临湖而立，景色宜人。设计以装饰主义的室内设计风格，打造出顶级现代文化时尚会所。转角处的名画、雕塑与艺术装置，烘托出独一无二的高尚品位。名师主理的中西餐饮、私密的雪茄房、专业的红酒窖、艺术品商店及高端会员俱乐部，体现出细腻的经营理念和人文关怀。

大间空间设计有限公司　/　江俊浩　/　李欣、侯秀佩、萧以颖

中国台湾新北市中和区中山路 / 2645m²

晶宴中和会馆

In order to have banquet space different from other general design framework and in accordance with stunning design thinking, the designer takes a highly visual touch way to interpret the spatial atmosphere. The use of stone, glass mosaic, shell board, half-silvered reflecting glass, crystal board, etc., makes the space more vivid.

为了有别于一般宴会空间设计框架并符合立体的设计思维，设计师采取高度视觉感官接触的方式诠释空间氛围。石材、玻璃马赛克、贝壳板、银半反射玻璃、水晶板等材质的使用让空间更加灵动立体。

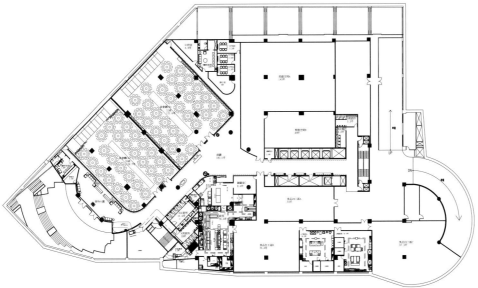

Food Space 77

Yakoya Kelapa **Gading**

Jakarta, Indonesia / 350m²

The design of the interior reflects the nature of traditional Japanese identity with a touch of the contemporary forms. Materials and design features such as the exposed plaster, clear stained timber, decorative bricks, treated bamboos and the combination of Japanese Sakura flower motifs dominate the whole spaces. Copper plates add final touches to the interior details, and light creates a soft ambiance in the whole space.

室内设计运用现代结构手法表现日本传统特征。材料和设计手法,如原始灰泥、褪色的木料、装饰砖块、加工过的竹子和日本樱花图案的结合,支配着整个空间。铜盘为室内细节作最后着色,灯光更为空间增添柔和气氛。

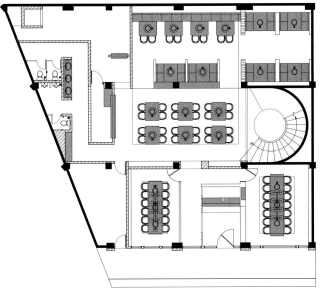
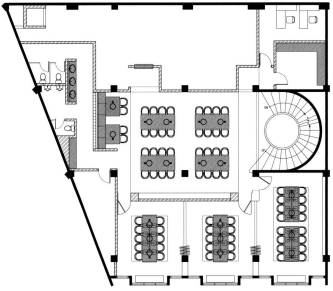

Food Space

无锡市上瑞元筑设计制作有限公司　/　孙黎明　/　陈凤磊、周怡冰

中国江苏省无锡市保利广场3楼 / 800m²

宴遇·乡水谣

The project is inspired by "Kangmei Love", a touching song reflecting romantic stories that are favored by people who love fashion. Therefore, the design makes the requests of mind and body of the prevailing and fashionable populations as the design core, and the space is mainly designed in blue color, boasting its intelligence, romance, elegance and vividness presented in the spatial effects.

本案设计灵感来自歌曲《康美之恋》——一个属于风尚阶层清新浪漫的美丽情愫。所以设计以风尚主流人群身心诉求为核心，以大面积的蓝色为色彩基调营造知性、浪漫、高雅、明快的时空感。

Food Space

宴遇·乡水谣

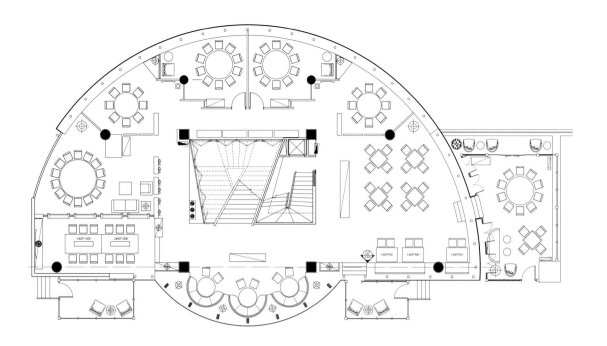

Neri&Hu Design and Research Office / Lyndon Neri & Rossana Hu / Dirk Weiblen

Shanghai, China / 150m²

House of **Wine**

Located on the top floor of the historic Three on the Bund building, House of Wine combines the intimacy of a wine cellar with the urbanity of a Shanghai rooftop bar. The whole space is covered with red clay bricks. The floor plan is split into two components, a bar area and an adjacent space that features more bistro-like seating. Inspired by the materials of wine barrels, the bar area uses wood and metal as the main materials. Wooden cabinets enclose various functional needs like a panoramic window framing the otherwise hidden view towards Pudong. The brick wall directly opposite the bar features various sized niches to allow a playful display of selected wine. A large sliding door divides the indoor bar space and the outdoor terrace. The wine cellar is inserted as a solitary object. Inside, wine shelves surround a central table for tasting events. The glass facade allows for a visual connection into the entire bar. The use of simple material palette of bricks, steel, and wood is not only a sensory experience but also an enjoyment of wine tasting.

House of Wine

坐落于极具历史感的外滩三号顶层的好吧，兼具了地下酒窖的隐秘性以及作为上海屋顶酒吧的都市感。整个空间被红色黏土砖包围。楼层分布分为两个部分：酒吧区域以及类似于小酒馆式的座椅设置。酒吧区域对于木头和金属材质的使用来源于酒桶材质的灵感启发。木质酒柜在达到多重功能需求的同时，如同一扇全景窗，构建出隐藏着的浦东美景。与酒吧相对的砖头墙面，开凿着不同尺寸的壁龛，随性地展示着特选的酒。一扇巨大的滑门将室内的酒馆空间和室外的露台分隔开来。酒窖被单独地置入。内部放置酒瓶的架子围绕着进行品酒活动的长桌。酒窖玻璃质的外表面将整个酒吧映入视野。砖、钢以及木材这些简单材料的混合运用，不仅是一场感官体验，更是一场品酒享受。

陕西陈维思设计顾问有限公司　/　陈维思　/　谢凯如、刘国平

中国陕西省西安市曲江新区 / 3000m²

唐华乐府

The project is located in Great Wild Goose Pagoda scenic zone. The design aims to transform a building (built in 1985) to a sophisticated restaurant fusing modern Chinese art performance. Utilizing its own height, volume and the surrounding environment, the design creates a dramatic atmosphere for all the diners through visual effects of the new materials to interpret traditional Chinese elements and culture.

唐华乐府

本案位于大雁塔景区，设计目标是将一个建于 1985 年的建筑改造为融合现代艺术演出的高雅餐厅。设计利用其自身高度、体积及周边环境优势，通过现代材料的视觉效果展现中国传统元素和文化，打造出有品位的用餐氛围。

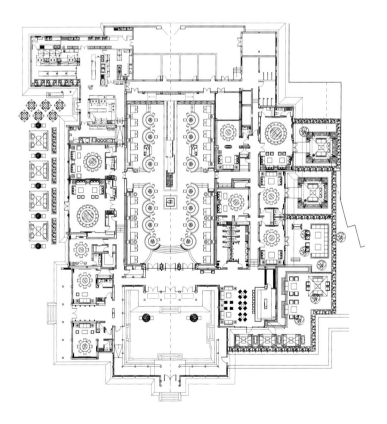

Food Space

Pollen **Street Social**

813 Pollen Street, London, UK / 570m²

Inspired from the meaning of the "social", the design fuses the architectural elements into a kind of social dining experience with practiced technique. The design for the restaurant's facade is indeed a nod towards the historic structures surrounding the site. A series of blackened bronze metal frames act as a stitching strategy, and a combination of transparent and translucent glass ensures visual continuity between diners and the life of the street beyond.

Within the ground floor spaces of the restaurant, the inviting atmosphere is reminiscent of entering the chef's own home, falling into familiar dialogue with an old friend. The design uses contemporary and abstracted technique to re-interpret the English classical details, such as the wood wainscot wall and the green glass P-Lamps for the bar, which retains a domestic ambiance while reflecting the elegance of the restaurant. Dramatic ceiling openings introduce plentiful light into the space, which echo jeweled pendant chandelier, attracting the sight of the customers in competition with delicious food. The Private Dining Room features wine fridges enveloping its perimeter, providing an enclosed yet visually open environment for intimate gatherings.

餐厅的设计理念来源于"社交"一词的涵义,以纯熟的技术,将建筑的元素融入一种社交的用餐体验。餐厅的外观设计,展现出其对餐厅周遭历史建筑的尊重。以一系列漆黑的青铜金属框架的植入,来取得与现有结构比例相协调的效果,玻璃与磨砂确保了视线的通透感和延续性,恰似引导着一场餐厅食客和街上行人的视觉互动。

餐厅一楼营造温馨的氛围,就像到了主厨的家,轻松的感觉像与一位老朋友聊天。设计以抽象手法重新解构英国古典细节,如木质壁板墙面及为吧台设计的绿色玻璃"P"灯,既保留了家的品味,又展现顶级餐厅的高雅。挑空的天花设计引进大量光线,辉映着光彩夺目的吊灯,与餐盘中美食竞相吸引着客人的目光。私人包房内,一组酒柜形成了隔断,在确保了私密性的同时也为私人社交活动提供了开阔的视觉空间。

Pollen Street Social

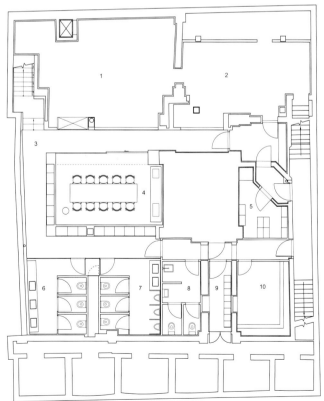

1. Main Kitchen
2. Main Kitchen
3. Corridor & Wine Cellar
4. Private Dining Room
5. Sommelier Room
6. Female Toilet
7. Male Toilet
8. Staff Toilet
9. Staff Changing Room
10. Office

HASSELL Ltd. / Scott Walker, Rebecca Trenorden

Yarra Lane, South Yarra, Victoria, Australia / 180m²

Yarra Lane **Precinct**

This project has three outlets, and each outlet is designed as if it is a unique entity, allowing the lane way to be characterised by the diversity of individual shops. Each shop front not only hints at the interior experience but also opens out, allowing food collaboration to spill into the lane way. The success of this project is based on not only the good operating activity of the restaurant and bar but also the increased people who are now using this lane as their active place.

Yarra Lane Precinct

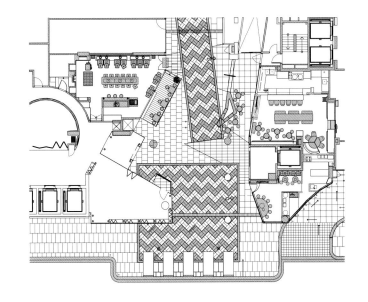

本案有三个出口，每个出口被打造得仿若一个独立体，通道也被个性化商店的多样性点缀得颇具特色。每个商店的正面不仅隐身于室内，同时也对外开放，允许食物集中进入通道。本项目的成功不仅在于餐厅和酒吧的生意兴隆，更在于增加的人流把这条通道作为他们活动的目的地。

Kubota Architects & Associates inc. / Shigeru Kubota

Galaxy Casino, Macau, China / 285m²

Mercedes-**Benz Connection**

This facility is a new style of brand information transmission hub, with Mercedes-Benz located within a café, restaurant, bar, and lounge. In order to activate communication among people, the concept of "creating interest" has been used as a theme. To allow the brand gain recognition as family, images and light are projected to create various scenes and landscapes, dramatically changing the space to create a new community. This approach aims to communicate the beauty of Mercedes-Benz in everyday scenery along with the pleasure and experience that comes with riding a car.

该设施是一个品牌信息传播中心的新样式，将奔驰汽车安置在咖啡厅、餐厅、酒吧及大堂内。为了促进人们交流，就有了本案"制造兴趣"的主题理念。为了使品牌被人熟知，重点应用形象和灯光来制造不同的场景和景观，戏剧性地将空间打造成一个新的社区。该方案旨在传达奔驰汽车在日常风景中的美感以及驾驶汽车带来的愉悦和体验。

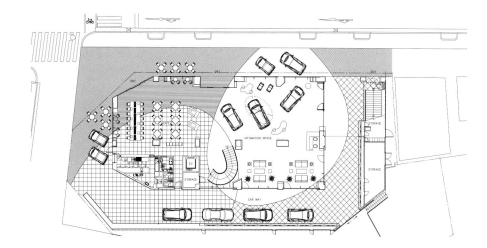

Mercedes-Benz Connection

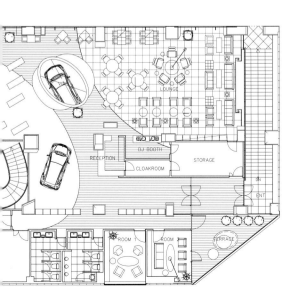

Food Space

4N Limited / Sinner Sin / Danny Ng

Hong Kong, China / 370m²

DIM SUM BAR

The interior design is inspired by the bustling fish market. The chosen materials depict the raw setting character, which perfectly defines the old and worn out fish market. Rustic copper plates, distressed pine timber, and metal plates are the main materials used in the design of the interior. The dimly ambiance creates a cozy and comfortable feeling for the diners.

设计灵感来源于繁忙的鱼市。精选的材料透着原汁原味，完美地诠释了鱼市的古老与破旧。粗糙的铜板、原色的松木、金属板材是设计的主要材料。朦胧的氛围带给就餐者安逸舒适感。

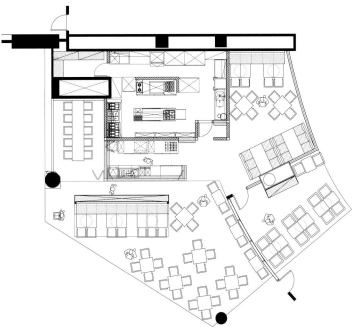

Liquid Interiors Limited / Rowena Gonzales / Leo Leung

48 Wyndham St. Central, Hong Kong, China / 81m²

En**vie**

This bar lounge, which was originally a black box, has been wholly renovated after a party, and becomes a welcoming place for lunch, tea time as well as a bar lounge. Inspired from the rhythm of music and poetry, the design features the long space with high ceiling in the flowing light and sculpts. The gold mirror at the end gives an optical illusion that the space continues toward the back. Some other elegant finishes such as a gold metal bar, glass tables and purple upholstery give the space a luxurious touch in contrast with the loft-like open ceiling and graffiti artwork paintings.

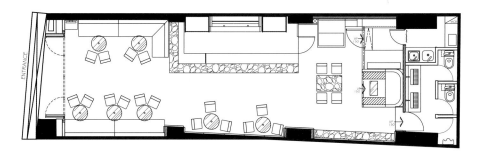

这间酒吧大堂，原来就像一个黑盒子，在一次社交聚会后，进行了一次全面整改，转变成一个就餐、品茗、小酌的好去处。源自诗歌音乐的韵律感，设计用流动的灯光及雕塑来装饰这间拥有高天花的长空间。尽头的金色镜子营造出一种错觉，仿佛空间一直向后延伸。金色金属吧台、玻璃桌子及紫色装饰物等典雅的装饰材料，在阁楼样式的开放天花和艺术涂鸦的对照下，呈现出一种奢华气派。

D&C.,Ltd. / Megumi Takahashi

Sapporo City, Hokkaido, Japan / 230m²

Orange Terrace

The Orange Terrace is located in the vibrant nightlife distinct of the heart of Sapporo. A walk down to the basement of a modern city tower will bring you on a journey through a fairy tale forest to a serene modern space which embodies the warmth, tenderness and sophistication in Hokkaido, where you will look forward to coming back to.

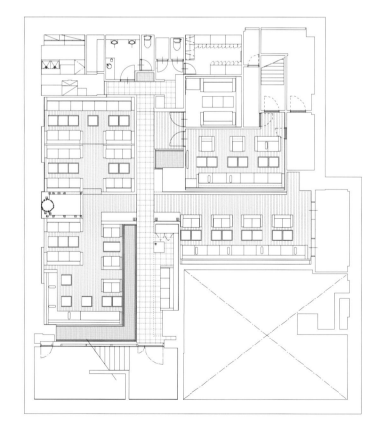

橘色梯田位于札幌中心的夜生活区。沿着一座现代城市塔的基座行走,你将步入从童话森林到宁静现代环境的美妙旅程,温暖,亲切,充满世情,令北海道成为人们流连忘返之地。

Orange Terrace

Food Space 119

Tsujimura Hisanobu Design Office / Tsujimura Hisanobu

Kyoto, Japan / 111.34m²

Gion NITI

The project boasts soft chair, solid teak wood counter, massive plaster ceiling, handmade Japanese paper wall, and linear glass floor, featuring a traditional style.

本案采用了柔软的椅子、实心柚木的柜台、大面积的石膏天花、手制的日本墙纸以及线形的玻璃地板，营造出一种传统风格。

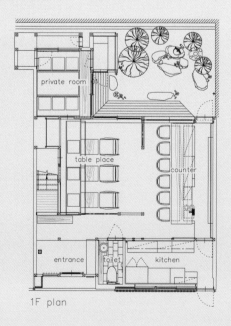

1F plan

Food Space

2F plan

Food Space 123

上海极思态室内装潢设计咨询有限公司 Zestyle Shanghai / Toh Yueh Loon / Adrian Chua

中国浙江省杭州万象城 5 楼 / 920m²

亚洲土地

The design uses creative screens, ceilings, lightings and artifacts to decorate the space, such as lighting and water jet jar made in garlic skins which is pro-environment element, the wave-like manner partition which is replica of the Great Wall of China feature. Yellow and Purple colors are known as the Royal Thai color and these colors are used on ceilings and wall respectively. The design aims to create a unique feel of Thai luxury and plush interior settings.

设计运用了创意屏风、天花、灯饰和传统文物来装饰整个空间，如环保元素——大蒜皮制成的灯饰和喷水缸，中国特色——长城造型的、呈现波浪式的屏风分区。黄色和紫色代表泰国皇家的颜色分别用在天花和墙壁。设计旨在创造一个不囿于传统泰国设计理念，而具有独特泰元素的豪华空间。

Food Space 125

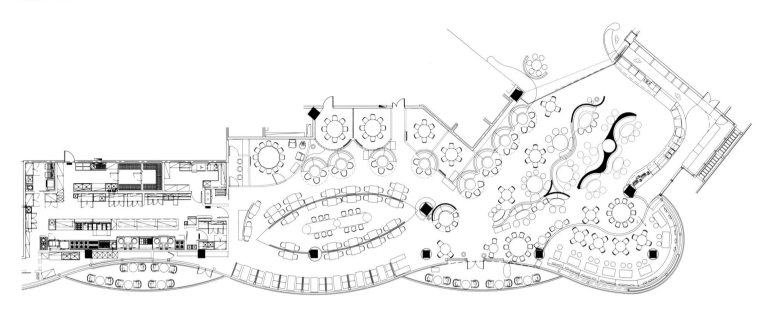

中国江苏省昆山市人民路89号 3F / 1440m²

叙品咖啡厅

The restaurant is well-located and possessed of roomy layout. The design adopts modern approach and combines with the interior landscaping decoration elements to fuse catering and culture, creating a tranquil and elegant atmosphere. At the entrance, the architectural walls with large-area wood block superposition provide a pure background for the environment, and combine with the cashier desk of bar counter and lighting fixtures perfectly, settling the modern simple style of the restaurant, while the use of gold foil facing heightens the spatial ennoblement.

本餐厅拥有优越的地理位置及宽敞的布局。设计用现代化的手法，结合室内景观装饰元素，将餐饮与文化融合在一起，营造出一种安静高雅的氛围。入口处大面积木块叠加的装饰墙为环境提供了一个很纯粹的背景，与吧台收银台以及灯饰完美结合，奠定了餐厅现代简约风格，适度使用金箔饰面，渲染出空间的尊贵感。

叙品咖啡厅

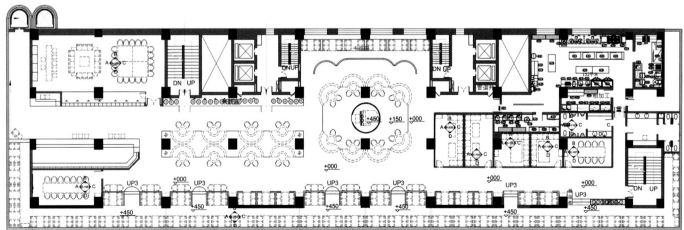

Food Space

百观设计 / 曹轩鋐、郑明智 / 潘婷、黄梅

中国北京市朝阳区远洋光华国际大厦 / 1200m²

北京辉哥火锅

As a high-end hot pot restaurant, "Huige" provides delicious top-level ingredients, where the design is originated. Imagine that the restaurant is a hunting resort and outside is a vast prairie and lush white birch forest, at dusk, lively voices and continuous laugh are heard and overflowing fragrance is smelt in the smoke, while people are sharing today's victory.

Food Space 133

北京辉哥火锅

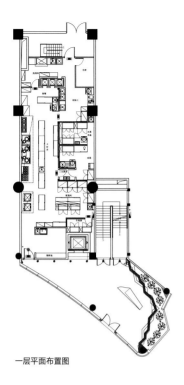

一层平面布置图

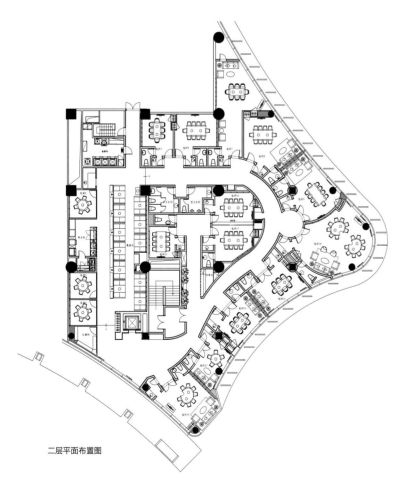

二层平面布置图

作为一间高档的火锅店,"辉哥"提供了鲜美顶级的食材,设计便从这里出发,想象餐厅是狩猎度假屋,屋外就是广阔的草原及茂盛的白桦林,黄昏时屋内人声鼎沸,欢笑不断,大家分享今日的战果,在炊烟之中已经闻到四溢的香味。

武汉朗荷室内设计有限公司 / 马先锋 / 胡耀伟、谢玉白

中国湖北省武汉市金银湖畔 / 1800m²

和宴

Based on the cultural aspiration of the owner and the high-end client orientation, the designer has a clear idea and expressing direction for the space to create the aesthetic pursuit and the visual experience. Common understanding and love of architectures with Chinese-Western culture in old Shanghai and Hankou during the period of the Republic of China, is the starting point of the design to demonstrate the aesthetic pursuit with a contemporary perspective. The originally uses of the layout, color tone, lighting, material and art furnishing depict a roll of elegant picture featured by grace, low-key and humanity.

基于业主的文化诉求和高端客户定位，设计师对于将要创造的空间美学追求和视觉体验有了清晰的思路和表达方向。对老上海、老汉口等民国时期具有中西文化色彩建筑的理解和喜爱，是设计的切入点，以当代视角剖析审美追求。运用布局、色调、灯光、材质以及艺术陈设等方面的匠心营造，描绘出一卷集雍容、低调、人文于一体的优雅画面。

和宴

江苏省海岳酒店设计顾问有限公司　/　姜湘岳　/　陆春丽、徐云春

中国江苏省南京市鼓楼区中山北路 1 号紫峰大厦 4F / 1504m²

南京紫轩餐饮会所

Zixuan Club, on the fourth floor of Zifeng Tower, the highest building in Jiangsu Province, is positioned at being the miracle of catering industry in Nanjing and the landmark of cuisine at Greenland-Square, servicing for elites.

Based on the orientation, the design adopts a creative thought of mixing modernism style and neoclassicism style instead of traditional Chinese or Continental style, endeavoring to create a kind of cultural atmosphere for people to immerse in and to feel the modern fashionable air.

南京紫轩餐饮会所

紫轩会所，位于江苏第一高楼——紫峰大厦的四层，定位于南京餐饮业奇葩、绿地广场的美食地标，倾情服务精英人士。

基于这一定位理念，设计抛开了非中即欧的传统思想，采取一种融合现代主义和新古典主义风格的设计创想，力求让人们沉浸在优雅的文化氛围中亦能品味到现代时尚气息。

Galaxy Casino, Macau, China / 285m²

Spice **Garden**

The design uses a combination of historical and cosmological references to plan a modern restaurant space within tight constraints. The venue is positioned on an internal arcade; the use of large glazed panels and horizontal louvers creates an external street like feeling, adding to the bazaar atmosphere, as well as a feeling of spaciousness within its tight floor area.

To suit the client request of a bright colorful interior, inspired by North India, the design uses intense, rich and earthly colors and materials, such as teak and ochre, punctuated by bright spice and flower colors. The aesthetic of the interior is an original combination of disparate inspirations from North India. Islamic geometry, a cosmological sun pattern, tall green bamboo, bright spices and an evening light blue, against rich earthly colors, come together to form a rich and lively interior design scheme.

Spice Garden

为了在紧凑的结构中打造一个现代的餐厅空间，设计参考了历史和宇宙方面的书籍。本案位于一座内部拱形建筑的上方，巨大光滑的面板以及与之平行的天窗，营造出一种街道感，增添几许市集氛围，让人在狭窄的空间中生出宽敞感来。

源自北印度的灵感，设计采用了热情奔放、鲜艳而世俗的色彩和材料，像柚木，赭色，点缀上亮色和鲜花色，满足客户想要一个色彩明亮的空间。北印度风情，伊斯兰图案，绝然不同的妙思的新颖结合，宇宙的太阳图案，修长的翠竹，明亮的调料，蓝色夜光灯和泾渭分明的世俗色彩的碰撞，构筑出一个丰富而鲜活的室内主题。

Inarc Design Hong Kong Limited / Terry Antony Spinolo / Fannie Bau, Sandy Lam

Happy Valley Clubhouse, Shan Kwong Road, Hong Kong, China / 2500m²

Derby Restaurant & Bar

Inarc Design won an international competition to provide the interior design for a phased renovation of the existing 8,000 square meters Clubhouse including the signature Derby Restaurant and Bar, where special spaces are created through the facilities including the entry vestibule echoing the stone wall patterns of old Italian farmhouses, the horseshoe bar clad with gold onyx, a lounge area with a special acoustic timber feature wall and a large spiral staircase framed by waterfalls leading to the intimate dining areas. Unique horse racing artwork was recently selected and installed in the Derby Restaurant and Bar to display recent winners of the annual Derby Cup Race.

建亚设计赢得了国际竞争，为香港赛马会跑马地现有8000平方米会所提供分阶段室内设计，包括标志性"打吡"餐厅及酒吧，整个设施均设有特别的空间，包括入口前的意大利农舍式石墙图案，酒吧的包金缟玛瑙，休息区的有声木材墙和一个大螺旋楼梯配以瀑布通往尊享用餐区。近期，独特的赛马艺术作品被选定及安装在"打吡"餐厅和酒吧，以便展示一年一度打吡杯比赛中的优胜者。

Derby Restaurant & Bar

Food Space

中国江苏省苏州开发区 / 450m²

苏州酒田日本料理

Based on modern design approach, the project adopts newly technology and materials to show the essential character of Japanese culture: purifying, subtlety and unadorned. The project takes the meaning of Japanese culture, but the use of modern element materials, such as steel structure doors of more than 3 meters, the walls and ceilings paved with a large area of complex-floor, and half-clarity black steel structure separate wall, representing the modernism spirit. The design makes the space interesting with the design approach of falseness and actuality.

本案用现代主义手法，采用新技术材料表现了日式文化的本质特征：精炼，微妙，朴实。苏州酒田虽然取意日式文化，但对充满现代主义元素的材料的大胆运用，如3米多高的钢结构移动门，大面积复合地板铺设的墙面及天花，半通透黑色钢构隔墙，则又透出现代主义精神。设计虚实互衬，显得空间趣意盎然。

苏州酒田日本料理

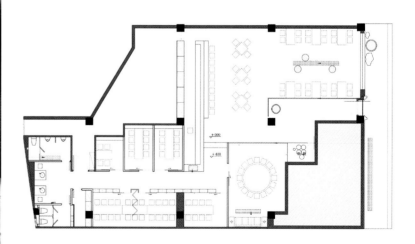

Shanghai, China / 200m²

OHA**NA**

Ohana blends the traditions of Japanese architecture, with contemporary Japanese culture to create a recognizable and traditional restaurant, which upon closer inspection reveals another slightly weirder layer.

Taking the ancient Japanese craft of handmade paper as a decoration strategy, a "comical" motif is applied to create a series of seasonal wallpapers that correspond to the changing environments; winter, summer, spring, autumn – blue, green, pink, orange respectively.

Amongst the graphic flowers (Ohana), godzilla, sumo wrestlers, ultraman etc., are intertwined and then applied to the restaurant ceiling and VIP rooms, creating an overall environment that is both diverse as it is similar.

Photography Tristan Chapuis

Ohana 是一家由多种日式传统建筑元素和其现代文化混搭而成的传统餐厅，仔细观察会发现其传统中又透出些许怪异，极具辨识度。

以旧时日本手工纸的制造工艺为装饰方案，一组滑稽图案被用于墙纸制作，以此表现四季变换的环境效果；冬蓝、夏绿、春粉、秋橙，一一映衬。

其中就有花朵（代表 Ohana）、哥斯拉、相扑运动员、奥特曼等图案，它们层层叠叠、错综复杂，置于餐厅和 VIP 包房的天花顶上，将整个用餐环境营造得多元而又相似。

深圳市汇博环境设计有限公司　/　曹成　/　宋伟、郑岷、曾更慈

中国陕西省延安市 / 3215m²

阿林鲍鱼延安店

The restaurant is located on the 1st to 3rd floors of the hotel. The whole 2nd floor is a dining-hall; the 1st floor accommodates the lobby, administrative section and kitchen; the 3rd floor is a courtyard tea bar. The materials for the main decorative covers are simple with a combination of leather and split glass, concise and snappy.

The added steel structure stairs in the lobby are spiral upwards without support, thus provide free circulated space behind the stairs and strengthen the feeling of sculpture of the stairs. The design reflects the abstract form of the flower on the ceiling of the entrance lobby and uses marble medallion on the ground of the main corridor on the 2nd floor to reflect its representational form, thus strengthens the interior stereovision. The three big compartments on the 2nd floor can be opened to hold small scale cocktail parties and dining-together. The decorations of the rooms are simple. The large cone-shaped chandeliers composed of plated glass balls array become the main sculpture, and the oil paintings portray the local plowing and weeding scene.

本餐厅坐落于酒店的一至三层。二层整层为餐厅；一楼为大堂、行政区及厨房；三楼采光厅被用来做中庭茶事。主要装饰面用材简单，用皮质和拼接玻璃结合，简洁干脆。

大堂的钢构楼梯是加建的，旋转向上中间却没有任何支撑，这样做既给楼梯后面预留了流通空间，又加强了楼梯的雕塑感。设计将抽象的花形体现在入口大堂的天花上，而具象的形式以大理石拼花作为载体用在了二楼主廊地面上，加强室内空间的层次感。二楼中部两侧各有三间大包间，可以打开连通，用来举行小型的酒会和聚餐。房间的装饰很简单，用电镀玻璃球体阵列组成的大型锥形吊灯作为主要雕塑，而作为色彩点缀的油画体现的则是当地的耕耘场景。

武汉市IEA装饰设计顾问有限公司　/　王治、范辉　/　谢华、高翔

中国湖北省宜昌市华美达大酒店4楼 / 2600m²

御膳皇庭中餐厅

The restaurant design aims at creating ceremonious luxurious ambience and valued feeling. The design adopts the combination of the traditional Chinese and Western style and mix match technique to impress the traditional culture as well as meet the modern esthetic need. The designer focuses on the spatial preciseness and symmetry, and every transitional space is set with opposite scenery, fully using the traditional space philosophy – scenes vary when you move. The massive use of exalted yellow and red color which represents festival ensures the luxurious keynote, whilst the proper interlude of the cold tone of blue and green makes the space more splendid. The skillful use of mirror offsets the shortage of daylight, and simultaneously continues the space.

御膳皇庭中餐厅

餐厅的设计旨在创造隆重奢华的氛围和尊贵的感受。设计风格上采用了中西结合和混搭手法，既要表达传统文化又要符合现代审美需求。设计师强调空间的严谨、对称，每一个转折空间都设置了对景，充分地运用了传统空间哲学中的移步换景。大量使用表示尊贵的黄色和代表喜庆的红色保证了奢华的基础，而蓝、绿等冷色调的适当穿插，让空间显得更华丽。设计师巧妙地使用镜面，弥补空间自然采光的不足，同时也起延续空间的作用。

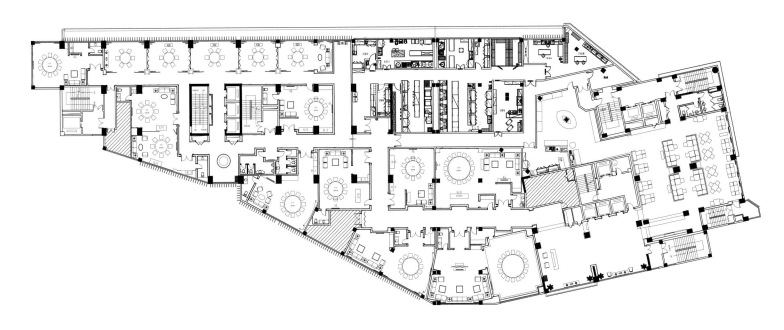

上海沈敏良室内设计有限公司　/　沈敏良　/　朱伟奇

中国上海市徐汇区天钥桥路214号哈尼美食广场45楼 / 969m²

Dolar **Shop**

The project is located on the fourth and fifth floors of the building, and the design idea is expressed by an umbrella which is inspired by the outdoor gazebo: nostalgia – modern ladies in cheongsam in old Shanghai with an umbrella; solicitude - lovers leaning together under an umbrella in the rains; vogue - the umbrella pattern itself with strong visual impact.

该案位于建筑楼层的4-5层，由室外的露台，从而联想到伞来表达整个设计思想：怀旧——十里洋场摩登女郎身穿旗袍执伞的婀娜身影；关怀——情人雨中借伞相依的场景；时尚——伞形图案本身就具有强烈的视觉冲击力。

Food Space

Dolar Shop

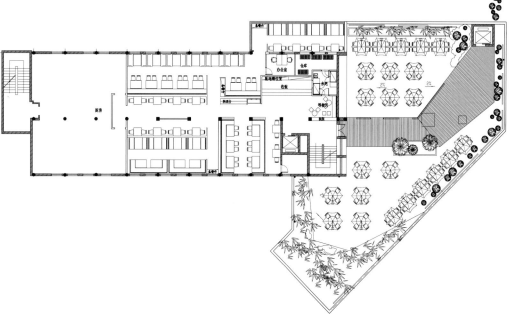

Food Space 171

中国上海闵行区都市路 5001 号仲盛世界商城内 / 893m²

Hannanshan **Korean BBQ**

Due to historical and geographical reason, Korean culture in ideology is always affected by Confucianism, which is composed of the design thought. Setting in the western culture and vogue elements, the design uses shading of the palisade in ornament technique to define the Zen realm, and with different styles of bamboo annotates the ideology of Confucianism.

由于历史及地理原因，韩文化在思想形态上一直受儒家思想的影响，从而奠定了本案的设计基调。以西方文化及时尚元素作为铺垫，设计用装饰手法中木栅条的明暗来演绎"禅"的有无境界，通过不同载体竹子的造型对儒家思想进行了最终的诠释。

Hannanshan Korean BBQ

中国山东省东营 / 2200m²

肴肴领先饮食会所

The restaurant is famous for healthy diet and suits for holding banquets. Using the modern Chinese style as theme, the design introduces decoration elements with classical Chinese charm into the whole space, as well as adds modern fashion elements. The plenty uses of local decoration materials satisfy the green and eco-friendly concept.

本案是以健康饮食及宴请文化为主题的餐饮空间。以现代中式风格作为设计主题，将具有中国古典韵味的装饰元素运用在整个空间中，并在其中加入时尚的现代元素。大量使用本地出产的装饰材料，符合绿色环保的新生活概念。

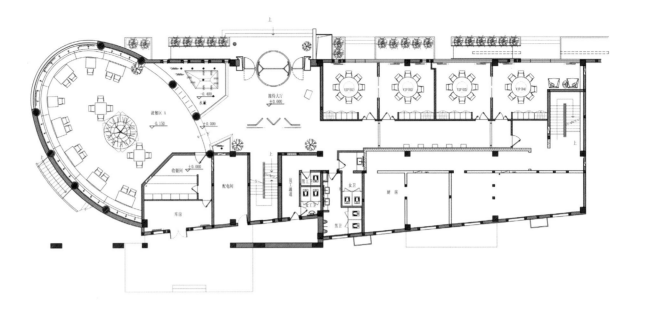

百观设计 / 曹轩鋐、郑明智 / 潘婷、陈静

中国上海静安区南京西路1266号506、601单元 / 1300m²

上海桃花源餐厅

Stunning grasses, falling flowers, orderly houses, as well as fertile fields, beautiful pools, mulberry and bamboo… how to re-interpret the beautiful scenery described in the *Land of Peach Blossoms*? As for indoors, the designers use modern materials to draw traditional lines, shining glass walls to enrich space level, taking stainless steel bamboo, well-proportioned courtyard style, water features and peach trees as the leading roles to reflect the serenity atmosphere in the space and the secluded charms of the paradise.

芳草鲜美，落英缤纷，房舍俨然，有良田、美池、桑、竹之属……《桃花源记》中所描写的美景要如何重新诠释？在室内采用现代的材质来勾绘传统的线条，用发光玻璃墙来丰富空间的层次，不锈钢的竹林，错落有致的院落造型，以水景、桃树作为主角，反射了空间中的宁静氛围，体现了桃花源的隐逸之美。

上海桃花源餐厅

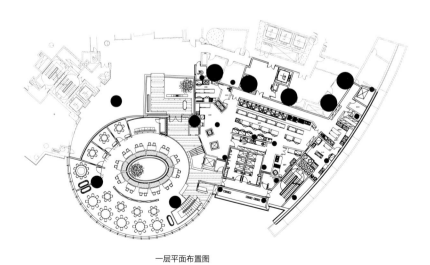

一层平面布置图

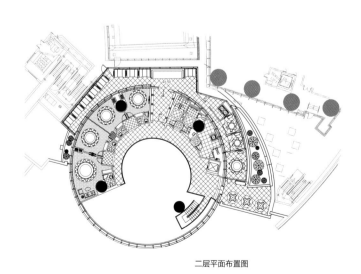

二层平面布置图

中国江苏省南京新街口国际金融中心 / 1200m²

港盈轩餐厅

The restaurant is located in a high-grade office building, with the elaborate design by the designer, bringing the nostalgic romance and visual experience of the perfect combination of Chinese and western culture for people. The colored drawing on the glass of partition, the wood-carving flowers on the ceiling, the European bronze droplight and green Chinese-style pumpkin lamp, make up a perceptual picture full of cultural connotation, which brings unparalleled experience for the guests.

该餐厅位于一栋高档写字楼内,经过设计师的精心设计,给人带来怀旧的浪漫气息和中西文化完美融合的视觉享受。隔断上的彩绘玻璃、顶面的木雕花、欧式的古铜吊灯、绿色的中式南瓜灯,构成了一幅充满文化内涵的感性画面,为宾客带来无与伦比的体验。

港盈轩餐厅

Metaphor / Martha Clarissa / Murni Khrisnawati

Jakarta, Indonesia / 150m²

Café **Regal**

The café aims to create a welcoming and yet a familiar interior space where the customers can feel like in their own kitchen. The materials, such as white washed timber strips, solid timber planks, painted bricks and copper light fixtures are chosen because they are rich in nature and typically used in traditional country kitchen setting. The open plan makes the space more welcoming with the counter as the main attraction to the whole space. A communal table placed in the middle of the space is used to attract a more casual crowd into the café.

Café Regal

这间咖啡厅旨在打造出一个既受欢迎又亲切的室内空间，顾客置身其中，仿佛身在自家厨房。选材上尽量采用自然界丰富并在传统乡村厨房中普遍应用的材料，如白色水洗木条、坚固的木板、彩绘砖以及铜质灯架。开放式设计让空间更受欢迎，柜台式的长桌尤具吸引力。陈列在空间中央的公用餐桌吸引更多闲散人群进入咖啡厅。

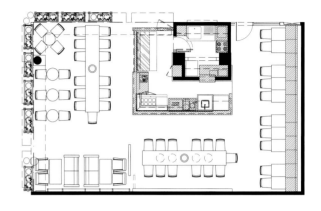

中国陕西省西安市 / 870m²

爱尚渔香主题餐厅

The project is a theme restaurant with fish as its featuring cuisine. The design, under the idea of taking fish as its theme, expresses the elements of fish, abstractly or concretely, through devices, wall painting, lighting, and partition, which makes new Chinese-style throughout furniture and details combined with simple but splendid space.

爱尚渔香主题餐厅

本案是以"鱼"为特色菜肴的主题餐厅。因此在把鱼作为主题的设计理念下，将鱼的元素通过装置、墙绘、灯具、隔断的形式，或抽象或具象地表现出来，结合简约大气的空间界面，将新中式风格贯穿在家具和细节之中。

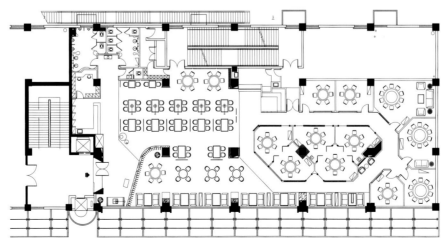

Food Space 195

Gandaria City, Jakarta, Indonesia / 210m²

Fish & Co

The design aims at creating a space that is revolved around the counters, where food is displayed and served. The counters as the main focal point of the space is designed with the intensive use of wood in 2 different tones, the off white and the dark timber finish. The materials are chosen and the ambiance is created to reflect the French bistro dining experience, where the customers can enjoy a simple afternoon tea or a light meal. Checkered tile flooring, sungkai wood and the baby blue mosaic tile behind the counter are the main materials that make up the vocabulary of the interior idea.

设计旨在打造一个围绕着柜台的空间，柜台正是陈列供应食物的地方。作为空间的焦点，柜台主要使用木材，设计成两种不同的风格，浅白和深黑的木料装饰。材料的选择、氛围的营造，让顾客在这间法式餐厅里乐享一个悠闲的下午茶或便餐时间。柜台后方多样的瓷砖地板，三开木及浅蓝马赛克作为主要材料，完美地定义了室内主题。

Fish & Co

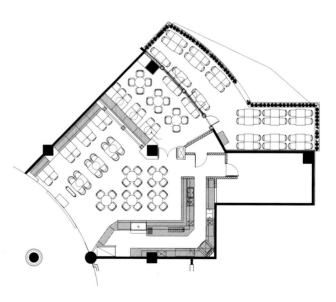